LP-Gas Tank Explosion Kills
Two Volunteer Firefighters
Carthage, Illinois

Reported by: Thomas H. Miller, P.E.
Michael W. Lackman

This is Report 120 of the Major Fires Investigation Project conducted by Varley-Campbell and Associates, Inc./TriData Corporation under contract EME-97-CO-0506 to the United States Fire Administration, Federal Emergency Management Agency.

Homeland Security

Department of Homeland Security
United States Fire Administration
National Fire Data Center

U.S. Fire Administration Fire Investigations Program

The United States Fire Administration develops reports on selected major fires throughout the country. The fires usually involve multiple deaths or a large loss of property. But the primary criterion for deciding to do a report is whether it will result in significant "lessons learned." In some cases these lessons bring to light new knowledge about fire — the effect of building construction or contents, human behavior in fire, etc. In other cases, the lessons are not new but are serious enough to highlight once again, with yet another fire tragedy report. In some cases, special reports are developed to discuss events, drills, or new technologies which are of interest to the fire service.

The reports are sent to fire magazines and are distributed at National and Regional fire meetings. The International Association of Fire Chiefs assists USFA in disseminating the findings throughout the fire service. On a continuing basis the reports are available on request from USFA; announcements of their availability are published widely in fire journals and newsletters.

This body of work provides detailed information on the nature of the fire problem for policymakers who must decide on allocations of resources between fire and other pressing problems, and within the fire service to improve codes and code enforcement, training, public fire education, building technology, and other related areas.

The Fire Administration, which has no regulatory authority, sends an experienced fire investigator into a community after a major incident only after having conferred with the local fire authorities to insure that USFA's assistance and presence would be supportive and would in no way interfere with any review of the incident they are themselves conducting. The intent is not to arrive during the event or even immediately after, but rather after the dust settles, so that a complete and objective review of all the important aspects of the incident can be made. Local authorities review USFA's report while it is in draft. The USFA investigator or team is available to local authorities should they wish to request technical assistance for their own investigation.

This report and its recommendations were developed by USFA staff and by Varley-Campbell & Associates, Inc. Miami and Chicago, its staff and consultants, who are under contract to assist the Fire Administration in carrying out the Fire Reports Program.

The United States Fire Administration greatly appreciates the cooperation received from Fire Chief Scott Carle of the Carthage Fire Department, Attorney John Pavlou and Special Agent Ted Anderson of the Office of State Fire Marshal, and Randy Boston of R.C. Boston Inc.

For additional copies of this report write to the United States Fire Administration, 16825 South Seton Avenue, Emmitsburg, Maryland 21727. The report and the photographs in color are available on the Administration's WEB page at http://www.usfa.dhs.gov/

U.S. Fire Administration
Mission Statement

As an entity of the Department of Homeland Security, the mission of the USFA is to reduce life and economic losses due to fire and related emergencies, through leadership, advocacy, coordination, and support. We serve the Nation independently, in coordination with other Federal agencies, and in partnership with fire protection and emergency service communities. With a commitment to excellence, we provide public education, training, technology, and data initiatives.

TABLE OF CONTENTS

LP-Gas Tank Explosion Kills
Two Volunteer Firefighters

Reported by: Thomas H. Miller, P.E.
Michael W. Lackman

Local Contacts: Chief Scott Carle
Carthage Fire Department
122 South Adams
Carthage, IL 62321

Attorney John Pavlou
Special Agent Ted Anderson
Office of the State Fire Marshal
1035 Stevenson Drive
Springfield, IL 62703

OVERVIEW

On Thursday October 2, 1997, two Carthage, Illinois Fire Department volunteer firefighters died, one was seriously injured, and another was injured when a horizontal liquefied petroleum gas (LP-Gas) tank BLEVE'd (Boiling Liquid Expanding Vapor Explosion). The rocketing tank struck them as they prepared to advance pre-connected hoselines from their high-pressure fog pumper about eight minutes after their arrival. The first fire company to arrive on the scene to a reported dryer fire found not only the grain dryer fire but also 30 to 40 foot intermittent fire plumes from the safety relief valves on two 1,000 gallon LP-Gas tanks, and a fully involved field tractor. Deciding on a direct attack, the first-in fire officer positioned the engine and firefighters behind a large grain silo 100 feet away from the burning tanks. But the engine's tailboard extended beyond the silo's vertical edge and was nearly inline with the end of one of the tanks. Three of the firefighters were advancing around the rear of the tailboard when a large tank section struck them.

The death of these firefighters demonstrates the serious hazards firefighters face when attacking LP-Gas tank fires. The need for first-in fire officers to be well trained in hazard and risk analysis and their ability to formulate effective action plans is critical to safe fireground operations. In many cases, LP-Gas incidents require the first-in officer to evacuate the area in anticipation of a tank failure rather than placing firefighters in position to try and prevent the failure. Analysis of the fireground factors present at the time of this incident indicate the chance of a successful direct attack to cool the tanks was unlikely given the severity of the fires.

1

The decision to make a direct attack is especially critical when an incident exceeds the capabilities of the initial crews to stop an escalating situation due to a lack of on-scene resources. In this incident, an immediately available and sustainable water supply for large volume hose streams was not rapidly obtainable. Even with an adequate water supply, firefighters may not have been able to apply the water with enough volume quickly enough to cool the tanks and relieve the excessive pressure before a failure occurred. The fire had been heating the tanks for over ten minutes and there was possible flame impingement on the top of the tanks.

When attack decisions are made, firefighters' position relative to the ends of the tank is critical. The Fire Chief knew that horizontal LP-Gas tanks fail and generally rocket in the direction parallel to the long axis of the tank and was working to avoid the ends. Firefighters located at the end of an LP-Gas tank are subject to being struck by the tank if the tank fails as this incident demonstrates. An additional concern over tank failures is the release and ignition of the liquefied gas, which quickly flashes to vapor and the energy released, can seriously burn or kill even fully protected firefighters. At this incident, the firefighters suffered no thermal injuries, but their intermediate stopping position in line with the tank's end reinforces the tactic that tanks should only be approached from the sides.

KEY ISSUES

Issues	Comments
LP-Gas Tank Location	The two tanks were located too close to buildings and too close to each other. Their location made application of cooling water difficult and created an exposure to the structure.
Tank Manifold	Connecting the two tanks by their liquid discharge to fill connections allowed liquid LP-Gas to back feed from the most exposed tank into the other, resulting in the hydrostatic failure of one tank.
Potential LP-Gas Tank Weakness	It is believed that a weak weld on a tank head to cylinder seam failed when it was subjected to excessive hydrostatic pressure. The condition would have been undetectable to the firefighters at the time.
Risk Assessment	The two 1,000 gallon LP-Gas tanks had been exposed to the burning grain dryer and likely gas being discharged from one of the tanks. The firefighters witnessed at least three intermittent releases from the pressure relief valves as they responded to the scene and prepared to attack the fire. The risk of tank failure exceeded the ability of the fire department to rapidly apply the volume of water necessary to cool the tanks.
Action Plan	Determination of the tank area exposed to heat and flame and time of exposure is critical to estimating the potential for BLEVE. Operating safety relief valves indicate high internal pressures that can lead to tank stress and possible failure. Operating relief valves are an indication that a direct attack to cool the tank surface is dangerous. When attempting such an attack, large volumes of water (typically master stream quantities) must immediately be directed upon the tank and firefighters must be positioned in safe locations.
Water Application Rate	The low flow rate, 60 gpm per hoseline high pressure pre-connects were unlikely to deliver the amount of cooling water needed to prevent the tanks from failing. In addition, these lines are difficult to leave unmanned. Flow rates of 250 to 500 gpm distributed over the entire tank surface are often recommended.
Attack Positioning	Attacks should always be positioned from the tank sides and from protected locations. At failure, horizontal tanks will rocket in the direction of its longest axis and leave a large fireball in its wake. Water and firefighter protective clothing will not protect firefighters from being physically injured or being burned by the ensuing fireball.

FIRE DEPARTMENT

The City of Carthage is located in west central Illinois, approximately 12 miles east of the Mississippi River. With a population of about 3,000, the city is the county seat for Hancock County. The city was laid out on a gridded street design where the downtown commercial buildings face a large town square. The Hancock County Courthouse occupies the center of the square.

Founded in 1877, the all-volunteer municipal fire department has twenty-eight (28) members who respond to approximately 50 fire alarms per year. Operating on an annual budget of approximately $40,000, the department is organized for structural fire suppression. The department provides no emergency medical services but assists the County's ambulance service when requested. The department is one of two municipal fire departments in the county and is active in the county's mutual-aid system that is composed of eleven fire protection districts and the two municipal fire departments.

Housed in a single-story three-bay station one street off of the city square, the fire department operates six pieces of fire apparatus. The primary city fire response vehicle is a 1995 1,250 gallon per minute triple combination pumper. In support is a 1948 eighty-five (85) foot aerial ladder, a 1976, 1,600 gallon tanker without a pump, a 1979 high-pressure, 120 gpm attack pumper with 1,000 gallon tank (used primarily for rural fire attack operations), and a rescue/equipment vehicle. Additional equipment includes a 1970 triple combination pumper with high-pressure capability, a 1959 pumper also with high-pressure capability, and a 1935 antique pumper. The department members' personal protective equipment consists of NFPA compliant structural firefighting protective clothing.

The department elects it's Fire Chief and two Assistant Chiefs each May and holds monthly business and training meetings on Tuesday evenings. The training is given by the department's training officer and consists of primarily Firefighter II level training with occasional pre-fire planning "walk through" of city buildings. Two (2) of the department's members are certified Firefighter III and four (4) certified Firefighter II from the Illinois State Fire Marshal's Division of Personnel Standards and Education. Additional training provided to its members was several University of Illinois' Fire Service Institute's training programs including an LP-Gas tank firefighting drill in 1995.

Although the department is a municipal fire department supported by municipal taxes, the department responds to rural fires in unincorporated property and other small towns immediately outside the city limits. The standard response to a reported structure fire would be either the 1,250 gpm pumper or high-pressure fog attack pumper with a minimum of three firefighters, the 1,600 gallon tanker, and the rescue squad (which carries the department's personal protective clothing). For rural structure fires, the first response vehicle is the high-pressure attack pumper because of its pre-connected high-pressure attack lines and its five seats. The second unit would be the rescue/equipment squad and then the 1,600-gallon tanker. The department's standard operating procedure requires a minimum of nine firefighters to respond to rural fire incidents. As rural property owners do not pay municipal taxes, the city bills the property owner $350 for the first hour and $250 an hour thereafter for any fire incident it responds to.

Emergency calls are received into and dispatched from the Hancock County Sheriffs 9-1-1 dispatch center. Paramedic level emergency medical service is provided by Hancock County operated ambulance service. Their vehicles are typically stationed at the local hospitals.

SITE DESCRIPTION

The incident occurred on a large farm located approximately 3/4 miles north of Burnside, Illinois and about ten miles north of the City of Carthage. The farm's principal buildings consisted of an occupied single-family frame home, a large wood frame shed, two metal clad pole sheds used for farm equipment storage and repair, and two large circular metal silos. A small, frame farm office building was located near the machine shop pole shed. Three unused wood frame barns were also on the property. (See Appendix A for diagram) The farm's principle crop was corn and soybeans.

No pressurized public water supply was immediately available at the site. The closest hydrant was located about 3/4 mile away in the town of Burnside and was supplied by a four-inch water main. A second public water supply was available seven miles away. The farm had two water ponds available for drafting operations.

The fire and subsequent explosion involved the farm's grain drying operation where harvested corn or soybeans was heated to reduce the moisture content before being stored. The grain dryer and supporting equipment was located to the west of the large wood frame shed and to the north of the west large metal grain silo. Although a portable unit with wheels, the continuous flow grain dryer had been at the location for many years. The dryer was principally constructed of metal with internal conveyors and passages that provide for fan forced heated air to dry the grain. The grain and the heating fuel were the primary combustibles in the dryer.

Mechanical power for the dryer's fans, conveyors, and grain movement equipment was provided by a power take off (P.T.O.) from a field tractor at the front (north) end of the dryer. Electric power for the dryer's controls was from an outdoor panel located southeast of the dryer near the pole shed between the two grain silos. (See Appendix B for diagram.)

The fuel supply for the dryer and the field tractor was provided from two, 1,000-gallon LP-Gas fuel tanks located immediately adjacent to the west wall of the destroyed wood shed. Distance between the shed and the tank closest to it was less than three feet and the tanks were less than five feet apart. The distance from the dryer to the tank closest to it was about 15 feet. Both tanks had integral steel legs, which rested on concrete pads. One tank had been at its location for many years and the second LP-Gas tank was added several years ago. At the time of the investigation, the identity of the original tank could not be determined. The two LP-Gas tanks, when full, provided about 36 hours of fuel for the dryer and tractor.

LP-Gas moved from the tanks to the dryer and tractor by means of a rubber LP-Gas hose. The hose connected the west tank to a tee fitting located near the front of the dryer. From the tee, fuel split to the dryer and the tractor. The hose reportedly lay on top of the ground between the tank and dryer. The hose was used for both vapor and liquid fuel transfer through different valved connections to the west tank. The dryer was indicated to start with LP-Gas vapor and then switched to liquid when warmed and under load. The tractor always used liquid and it did have an integral fuel tank that should have allowed the tractor to operate while the hose contained vapor.

At the west tank the LP-Gas hose was also connected to a tee. This fitting allowed the line to be attached to both the liquid withdrawal and vapor withdrawal connections on the tank. By means of manual valves, either vapor or liquid could be sent to the dryer and tractor.

When either LP-Gas tank emptied, the hose and fittings would be disconnected from the empty tank and attached to the full one. The dryer would have to be restarted and potentially any trapped air removed from the fuel hose. About two weeks before the incident, the delivery driver for the LP-Gas supplier suggested a means to manifold the two tanks together and eliminate the hose transfer process.

The driver suggested attaching a liquid transfer hose, used for filling LP-Gas fueled field tractors, to the liquid withdrawal connection on the top of the east tank and to the liquid fill connection on the west tank. No similar vapor space interconnection between the two tanks was identified during investigation. The liquid transfer hose did have a manual valve at one end but there was no indication of check valves, relief valve, or excess flow valves in the hose.

The tank manifold configuration worked satisfactorily and the two tanks were filled several times during the two-week period. Prior to the explosion, roughly 35,000 bushels of grain had been dried during the current harvest season without mishap. Both tanks had been filled in the morning of the incident and had been in operation for about seven hours without a problem.

The grain drying process began with grain brought from the field and gravity discharged into a field tractor PTO driven auger that deposited the grain into the wet bin. The wet bin was a vertical metal storage unit with a conical shaped bottom that acted as a surge bin for the dryer. It provided a continuous feed to the dryer while grain trucks from the field were changed.

From the wet bin, the grain discharged from the bottom by gravity into an electric motor driven auger, which discharged, at the top front of the dryer. Once inside the dryer, grain movement was provided by the LP-Gas fueled tractor, which mechanically powered the dryer. From the dryer, the grain discharged into another electric motor powered auger, which elevated the grain into one of the two storage silos. The unloading of the trucks from the field and discharge of grain into the wet bin was supervised. Once running and adjusted, the dryer did not require constant supervision. The dryer was started each morning and was stopped in the evening.

FIRE INCIDENT DESCRIPTION

Corn from the field harvest was being unloaded into the wet bin under supervision and the dryer was operating. Weather conditions at the time were a temperature of about 80 to 85 degrees Fahrenheit and light southwest winds of 5 mph. A loud "poof" was heard from the dryer followed shortly by a noticeable change in operating sound. Besides attracting the attention of the person unloading the corn, the noise was heard by four people working on a combine in the pole building machine shed. The unloading of corn from the field truck was stopped and the tractor PTO powered auger into the wet bin was also shut down. Fire was observed coming from the front, upper east half of the dryer. Shortly after the "poof", the LP-Gas fueled tractor, which powered the dryer, sped up. The speed change indicated that the dryer load had been lost likely due to drive belt failure on the dryer. In addition to the person unloading the corn, one of the people working on the combine came to investigate the "poof". After hearing the word fire being shouted, one of the farm's owners also left the combine to investigate.

The owner ran to the farm office and called the fire department via the county's 9-1-1 telephone system. At the same time, the person unloading the corn moved the field truck and then turned off the electric power to the dryer and auger conveyors from inside the pole shed located between the two grain silos. From inside the shed, he reported seeing the fire burning at the two LP-Gas tanks and venting from the tanks. He also reported that the fire had spread to the tractor at the front of the dryer. The tractor continued to run for a period after the fire spread to it. No one reported going to the LP-Gas tanks to shutoff the supply hose or the manifold hose.

Another witness, who arrived after the fire was reported to the fire department but before their arrival, indicated that the fire was burning on the front and east sides of the dryer and on top of the west LP-Gas tank. He also believed that liquid LP-Gas was being discharged from the liquid connection on the west tank. The liquid was spraying onto the east side of the dryer. The witness moved to a position near the farm office and reported that the fire burned violently at times and then the intensity would reduce dramatically. However, the flames never ceased entirely while he watched. He believed that the flames were concentrated on the west side of the west tank around and to the north of the tank's top connections. This area would be where the supply hose to the dryer and tractor would have passed. While he watched, a pressure relief valve began to operate intermittently producing a loud noise and flames 40 to 50 feet high. He also observed that the flames had spread to the top of the east tank, which could have been the result of the liquid hose failure between the two tanks.

The owner, after reporting the fire, went back to the machine shop and moved the combine from the building into the soybean field south of that building. Returning to a position near the farm office, he observed that the fire's intensity had doubled or tripled in the time he moved the combine. He then decided to contact the Carthage Fire Department directly rather than through the county 9-1-1 center. His purpose was to convey the worsening fire conditions and to suggest a neighboring fire department be alerted to respond for assistance. After talking with someone at the Carthage Fire station, he exited the office and observed that the east rear tire on the tractor powering the dryer was burning and that the tractor was still running.

The owner then proceeded to move a field tractor parked between the large wood frame shed and the east grain silo. To reach the tractor, he went around the south and east side of the silo. After parking the tractor to the south of the east silo, the owner ran to the electric power panel near the dryer and further shut down power in the area. The west tank vented as he was completing this task and the tractor north of the dryer was fully involved.

The tank or tanks vented several more times between then and the time when the first fire units arrived. Determination of which tank vented each time could not reliably be identified. The owner met the first engine company and informed them that two, 1,000 gallon LP-Gas tanks were involved and that the tanks had been filled that morning.

FIRE DEPARTMENT RESPONSE

About 4:39 p.m., a telephone report of the fire was received at Hancock County Sheriff's 9-1-1 dispatch center for a "dryer fire" in the town of Burnside, Illinois. The Carthage Fire Department was dispatched shortly after. Their initial response was the department's routine rural response assignment to what was thought to be a clothes dryer inside a house. The responding equipment was the high-pressure fog pumper (Engine 11), a 1,600 gallon tanker (Tanker 13), and the rescue truck (Rescue 10). The personnel on the first responding unit, Engine 11, consisted of the Fire Chief (company officer), Apparatus Operator, and a Firefighter. Immediately following was Rescue 10 with three firefighters. Tanker 13 followed shortly after with three firefighters. While enroute, the firefighters received a radio report from the County dispatch center that the fire involved LP-Gas tanks. The firefighters witnessed several openings of LP-Gas tank pressure relief valves from several miles away. Mutual aid was requested at this time from Dallas City/Colusa Fire Department for their 3,000-gallon tanker and from the LaHarpe Fire Department for an engine and a tanker.

Arriving at 4:48 p.m., Engine 11 and Rescue 10 pulled into the south entrance of the circular driveway, parking next to the farm office facing to the southeast. The farm's owner met the Fire Chief in the driveway and informed him that two 1,000- gallon LP-Gas tanks were involved. The owner's father also advised the firefighters that nothing involved in the fire was worth taking too big a risk and to not take any chances. The Fire Chief reported that from this position, he was able to see the tank ends, the grain dryer, and field tractor. The dryer and tractor were fully involved and the fire was burning at the two tanks. The Fire Chief and firefighters went to Rescue 10 removed and donned their protective turnout clothing.

A safety relief valve was operating intermittently and the exposure fires ignited the discharged vapors. Witness statements are not consistent regarding which tank's relief valve operated before the explosion. Before the fire department's arrival, witnesses indicated that the west tank's relief valve discharged several times. They also admit that the east tank's relief valve may have also operated because it was difficult to identify the exact tank. The fire chief believes that he observed the east tank venting at least twice before the explosion. The burning plumes were igniting the side of a wood frame shed next to the tanks.

Asking about the contents of the shed, the owner said that it only contained a few tires and some hay. The Fire Chief walked closer to the fires to better view the situation and develop a plan of attack. Concerned that the west tank was angled towards their position, he noted the east silo (Appendix A) would offer some protection from the tanks and would be a better operating position. He walked south of the west silo where he also noted the doors on both sides of the pole shed were open allowing a better view of the tanks and fires.

Returning to the engine, the Fire Chief found the firefighters had already pulled and advanced a pre-connected high-pressure 1-1/4-inch handline toward the fires. He ordered Engine 11 to relocate to the south side of the east grain silo and for Tanker 13 to establish a water supply at that site. Because a pre-connect had already been pulled, the Fire Chief and firefighters had to pick up the hoseline and walk it along the left side and rear of the moving engine to the new position.

Satisfied with the new position, the Fire Chief, who was walking at the left front of the engine, dropped the hoseline as the pump operator engaged the pump. Simultaneously, the two firefighters, who were walking the hoseline behind the engine, continued to move toward the Fire Chief. The rear of the engine was not completely behind the silo and the tailboard was almost in line with the long axis of the east tank. Because the burning dryer, tractor, and tanks were visible from the rear of the engine, the two firefighters, and another not involved with the movement of the engine, likely paused to observe the fire scene through the open doors of the pole shed just as the east tank BLEVE'd.

The tank separated at the weld seam where the north domed head was attached to the long cylinder shaped body. The tank head was broken into two pieces ("clam shelled") and the pieces traveled north and northeast into a brush and tree covered ravine area about 600 to 650 feet away. The balance of the tank rocketed to the south in a very shallow climb through the pole shed coming to rest nearly 1,000 feet away. (See Appendix C)

The tank struck several objects as it traveled south including three Carthage firefighters. The wood shed's six foot high concrete foundation was shattered along the west side from the explosion and the structure was destroyed by the fire. The west LP-Gas tank was thrown into the air passing over the grain dryer and wet grain bin, landing nearly upside down near the tractor which powered the auger that filled the wet bin from arriving trucks. The tank was discharging burning LP-Gas and the tractor caught fire.

The rocketing tank traveled through the pole shed as it proceeded to the south striking two door posts and a pipe rack in the shed. The wood 6-inch by 6-inch northeast door post was torn out from about one foot above the ground to about five feet above the ground. The tank then struck a glancing blow to a large steel constructed pipe rack inside the shed. The pipe rack is believed to have slightly altered the direction of travel causing the tank to turn slightly and out the open door on the south side of the shed. The wood 6 inch by 6 inch southeast door post was splintered from approximately five feet above the ground to just under ten feet.

Immediately on the other side of the shed's southeast doorpost stood the three firefighters at the rear of the Engine 11. Victim #1 was standing at the left corner of the tailboard and was knocked approximately 50 to 75 feet south. Victim #2, who had been standing to the left of Victim #1, was knocked approximately 130 feet to the south and into the soybean field. Both firefighters received severe traumatic injuries and died immediately as a result. The surviving victim, from the rear of the engine, had been standing behind Victim #1 and fell a few feet away. His injuries were serious and he was air lifted from the scene.

The tank did not strike Engine 11 although the apparatus was physically damaged. The damage consisted of some equipment mounted on the tailboard and the lower sections of the driver's side mounted ground ladders. After striking the firefighters, the tank continued south over the parked combine until it struck the ground the first time approximately 400 feet away. It continued to tumble and skip for another 600 feet through a soybean field, coming to rest approximately 1,000 feet away from its original position.

The fire chief was thrown to the ground and injured by the force of the explosion or from being struck by a flying object. He was able to request additional mutual aid assistance from the Terre Haute Fire Department, Crop Production Company, and for ambulances. After helping attend to the injured firefighters, he was also transported by ambulance to the hospital. The assistant chief arrived after the BLEVE and took command of the incident. He immediately began an accountability check of the onscene firefighters and farm workers. A telephone call to the Carthage fire station was made for names of responding firefighters. All firefighters and farm workers were accounted for at the end of the process.

On arrival, Tanker 13 set up its 3,000-gallon drop tank off the driveway north of the farm office. Engine 14 arrived and positioned to draft out of the drop tank and to direct its pre-connected deluge gun onto the still burning west tank's position. The tank was discharging burning liquid and vapor from the connections at the top of the tank. In addition, the field tractor, which powered the auger, caught fire from being sprayed with the burning LP-Gas.

Dallas City/Colusa's 3,000-gallon tanker shuttled water to fill the dump tank. La Harpe sent an engine and a tanker. The engine led out to a pond located east of the farm buildings and attacked the well-involved wood frame shed. Their tanker assisted in the water shuttle operation. Terre Haute also sent an engine and tanker. Terre Haute firefighters assisted with the shed fire and the tanker participated in the water shuttle. Crop Production Company (private business) supplied a field tanker with two 1,000-gallon tanks, normally used to fill farm equipment, to shuttle water. Hamalton Fire Department filled the Carthage station with an engine company. (See Appendix D)

The tanker shuttle provided Engine 14 with enough water to cool the LP-Gas tank allowing it to burn out, and to suppress the fire in the field tractor. Until LaHarpe's engine established a drafting operation from the pond, water supply was a problem on the east side of the fire scene. The fires were confined to the grain dryer, two tractors, and the wood frame shed. Engine 11 did not participate in the fire suppression operations.

In addition to the fires at the farm buildings, a large field fire occurred in a combined (harvested) soybean field about 700 to 800 feet north of the LP-Gas tank position. Although the burned area was searched for an ignition source, nothing could be identified as a cause for the fire. The field was in line with the long axis of the BLEVE'd tank and on the opposite side of the ravine where the broken tank head was found. There was no fire in the ravine and the field fire did not occur until after the explosion.

The Carthage Fire Department report indicated that units returned from the incident at 9:12 p.m. that evening. However, the La Harpe Fire Department provided scene security over night since the investigation of the incident had not concluded. Mutual aid departments provided coverage for Carthage Fire Department alarms from this point until after the funeral services for the fallen fire-fighters on the following Tuesday, October 7, 1997.

BUILDING AND FIRE PREVENTION CODES

The farm is located in Hancock County, which would have code review and enforcement authority. In addition, the State of Illinois and the Office of State Fire Marshal have enacted laws and regulations for liquefied petroleum gas storage and use. The state regulations acknowledge that compliance with National Fire Protection Association (NFPA) Standards shall be accepted as compliance with state regulations.

The analysis that follows makes use of NFPA Standard No. 58, *Standard for the Storage and Handling of Liquefied Petroleum Gases* and National Propane Gas Association (NPGA) documents #500-93, *Safe Use of Propane with Crop Dryers* and NPGA #613-92, *Guidelines for Manifolding Liquid Withdrawal ASME Containers not Exceeding 2000 Gallon W.C. at Construction Sites.* NFPA Standard No. 54, National Fuel Gas Code does not apply to crop dryers or to the LP-Gas fueled tractor.

According to NFPA Standard No. 58, the two 1,000 gallon capacity tanks should have been located at least 25 feet from any building. The purpose of this distance is to provide exposure protection for a fire at either the tank or the building. A building fire would not immediately threaten the LP-Gas tank and a fire at the LP-Gas tank connection or relief valve would not immediately start the building on fire. Another reason for the separation distance is to minimize the potential for escaping heavier than air LP-Gas to enter the building potentially finding an ignition source.

The tanks should have also been positioned with at least three feet of space between the two tanks. This space allows access for water streams to cool the tank surfaces and for personnel to reach the valves on the top of the tank. Another important separation distance is between a gas discharge point on the tank and potential ignition sources such as the grain dryer and tractor. NFPA Standard No. 58 indicates that at least ten feet should be provided. Potential gas discharge points are the fill connections, the pressure relief valve, and hydrostatic relief on liquid piping. The NPGA also recommends that grain dryers be shut down during LP-Gas tank filling unless the distance between the dryer and the tank is over 50 feet.

NFPA Standard No. 58 limits flexible hoses to 36-inch maximum length. The LP-Gas supply hose from the tank to the dryer exceeded the maximum and should have been replaced with Schedule 80 pipe and high-pressure fittings. The heavy piping and fittings are needed because the line contains liquid LP-Gas and the operating pressure could exceed 250 psi. The piping between the tank and the dryer/tractor would also need to be protected from mechanical damage. The pipe should be buried, with adequate corrosion protection, or suspended on supports above ground. The liquid supply piping would also need an excess flow valve, shut-off valve, and a hydrostatic relief valve. The excess flow valve is to stop liquid discharge if the line fails catastrophically. The shut-off valve is the manual means to stop the LP-Gas flow or to isolate the line for other work. The hydrostatic relief valve is intended to protect the line if the trapped liquid expands as its temperature increases. The excess flow valve and hydrostatic relief valve were needed on the liquid line regardless of the line's construction material.

To comply with NFPA Standard No. 58, the pipe lines used to manifold the two 1,000 gallon tanks should have been constructed of Schedule 80 pipe with extra heavy fittings because the pipe could experience pressures over 250 psi during normal operation. The process of manifolding the two tanks involves separate connections between the vapor space and the liquid space in each tank. Two separate pipes are needed to ensure that the liquid level in each tank remains balanced as internal pressure is equalized in the two tanks. The liquid manifold line could have made use of bottom outlet connections on each tank. Alternately, tanks manufactured after July 1, 1961 have liquid top outlets, which could have been used.

The liquid manifold piping would need an excess flow valve and manual valve at each tank and a hydrostatic relief valve in the line because of the possibility of trapped liquid. The vapor manifold line would also need an excess flow and manual valve at each tank.

The remains of various LP-Gas hose connections were gathered after the fire. A number of the fittings used to attach the hose were not industry approved for use with LP-Gas hose. Substandard hose fittings included stainless steel bands with adjustable screws. It is suspected that these connections did not have sufficient pressure rating for the LP-Gas liquid service.

The omission of a vapor manifold connection between the two tanks did not allow the liquid levels to equalize or balance in the two tanks. As a result, the fire heated the west tank and pushed liquid into the east tank until it was 100% liquid filled. The east tank then failed at its weakest point, the weld seam. The east tank's relief valve may have been unable to reduce the pressure fast enough or failed to operate effectively.

INVESTIGATION

A Special Agent from the Office of the Illinois State Fire Marshal was the lead investigator. He was assisted by members of the Illinois State Police; Hancock County Sheriff's Office, Hancock County Coroner, Illinois Fire Service Institute, and an independent engineering consultant. In addition to these agencies, various private cause and origin investigators and experts were permitted access to the site and equipment under supervision of the lead investigator. The objective was to identify the cause of the fire and to determine why the tank BLEVE'd.

Scene documentation began that evening and continued into the next two days. Photographs and videotape were taken at ground level and from a state helicopter. The aerial views provided an excellent means to describe the path that the east tank traveled.

The two LP-Gas tanks, parts of the LP-Gas hose fittings that were not consumed in the fire, and the two pieces of the domed tank head were transported to a secure location. The portable dryer and the LP-Gas fueled tractor, which powered the dryer, remained at the site. The extent of damage and size made their movement impractical.

The east LP-Gas tank was moved from the field where it came to rest with a tank carrier. The north head separated from the rest of the tank at the weld seam. The metal at the separation seam (fracture surface) was very smooth on both the tank and on the pieces of the head. The surface was almost like a grinder had passed over it. There were no chevrons[1] on either the tank or the head's fracture surfaces. The paint on the south end of the tank appeared blistered while the paint at the north end was unmarked.

At the time of the scene visit, the roll forming of the metal for the domed head to the tank cylinder weld seam was being reviewed for proper shape. In addition, the position and penetration of the weld bead at this seam was being analyzed. Both were possible contributing causes for the tank failure. Diagrams of the formed metal can be found in Appendix E and photograph of the weld seam in Appendix F.

The inside of the tank was clean with no visible corrosion or pitting. The liquid LP-Gas pickup tube had detached from the liquid withdrawal connection and was found near the tank's final resting-place. The tank did receive several dents and scrapes caused by the tank striking the ground and other objects. In addition, the tank's south domed head had a hole about 1/2-inch in diameter punched through the metal. Marks inside the hole suggested that it had been threaded. A steel rod, threaded on one end, was found at the third or fourth tank shell impact with the ground. The rod matched the hole in the head and was traced to the metal pipe rack that the tank struck in the pole shed. The threaded rod was a support piece for the pipe rack.

The west LP-Gas tank was also moved with a tank carrier to a secure storage location. At the time of the site visit, the original position and direction that the tank was facing prior to the incident had not been determined. It is hoped that the hinge connection for attachment of the protective valve cover may be used to determine which end of the tank faced north. (See Appendix F for photograph of the tank.) Around the hinge connection, a two to three inch high blister in the tank shell had formed over an area about 2 feet long by 1-1/2 feet wide roughly on top of the tank. The typical cause of tank metal deforming in this fashion is the direct flame impingement from burning LP-Gas on the tank shell in the vapor space. Allowed to continue, it is likely that this tank would have also BLEVE'd from the stretched and weakened metal.

The Office of the Illinois State Fire Marshal contracted with an independent engineering consultant to assist with the identification of the cause of the original fire. The consultant's report described two possible ignition sequences:

> Sequence 1: The noise from the dryer was caused by the failure of drive belts to the blower. The loss of the fan function allowed the dryer burners to flash out and consume the liquid propane lines and ignite the released propane vapors. The heat from the propane fire exposed the two 1,000 gallon LP-Gas tanks with the closer tank receiving the most heat. The heat increased the vapor level, which forced liquid propane into the other tank.

> Sequence 2: Failure of the flexible LP-Gas hose lines in the area of the tractor and dryer caused a flame roll-out from the dryer. The release of propane from the hose line resulted in the exposure of the two 1,000 gallon tanks to excessive heat.

---

[1] Chevrons are small v-shaped signs on a metal surface. They can be used to identify the point where a metal tear began or where a crack or fracture in the metal might have began. Protection of fracture surfaces from mechanical damage and corrosion or oxidation is very important for future metallurgical analysis. A coating of metallurgical oil is preferred but a lightweight lubricating oil can be used. Even motor oil thinned about 50% with mineral spirits has been used on fracture surfaces to delay corrosion.

LESSONS LEARNED

1. **The position of firefighters and apparatus should avoid tank ends, which are in-line with the long axis of the tank.**

   When LP-Gas tanks BLEVE, the energy released will often propel parts of the tank in directions parallel with the long axis. Approach and firefighting positions should be perpendicular to the tank's long axis to minimize the risk to firefighters. The Fire Chief's plan was to position the engine behind the large grain silo to provide shielding and position handlines between the grain silo and the animal shed to the east. However, Engine 11 was stopped before it was completely behind the silo, leaving the back of the engine exposed. Although the next step in the plan was not executed, the nozzle positions would have avoided the tank ends. Options for nozzle placement to wet the entire surface of both tanks were limited because of their location. The wood frame shed obstructed access to the entire east side of the LP-Gas tanks. The burning grain dryer and tractor restricted access from the west.

2. **To effectively cool LP-Gas tanks exposed to a fire and absorb the heat energy from flames impinging directly on the tank shell, a substantial water application rate is required.**

   For exposure fires, the entire tank surface must be kept wet to absorb the radiant heat energy reaching the vessel. When flames are impinging on the tank, a water stream must be constantly applied at the contact point to prevent the metal from weakening and thinning out. Water flow rates of 250 to 500 gpm for the two 1,000 gallon tanks would be a recommended minimum. The high-pressure, low flow rate, pre-connected lines would not provide the necessary amount of water on the tank. Furthermore, hose lines that can be placed into operation and then left unattended are preferred over hoselines that need to be manned. It is unlikely that the pre-connected hose lines could be left operating unattended.

3. **The more time the LP-Gas tank is exposed to fire conditions, especially to fire impinging on the metal shell, the greater the risk of BLEVE.**

   The structural integrity - usually defined as the ability to carry the load - of anything (floor, roof, building, pressure tank, etc.) exposed to fire usually deteriorates in accordance with the length of exposure time. Long delays in beginning the application of water (cooling and suppression) from the start of fire exposure, should raise warnings about possible structural failure. In this incident, the operating pressure relief valves are indications that the LP-Gas tanks are under stress and thereby increasing the prospect of catastrophic failure (BLEVE). Long fire exposure times and operating pressure relief valves should indicate that the potential for tank failure is imminent. The incident commander should concentrate on evacuating the area parallel to the long axis of the tank and limiting exposure to firefighters and bystanders. The minimal life exposure should be a significant factor in the risk benefit analysis.

4. **Tactical plans should anticipate that "portable" and "temporary" installations are not in full compliance with fire and safety codes.**

   Compliance with fire and safety codes cannot be assured for equipment installed on a "temporary" or "portable" basis. Plan review and inspection of these installations is often omitted even if officials are aware of their presence. Farms and construction sites are typical locations of "portable" and "temporary" equipment installations. Fireground operations and tactics should contemplate that fire and safety code compliance is likely to be incomplete. Fire conditions, spread, and reaction may be different than your experiences from drills and previous incidents. The long LP-Gas hoses and the noncomplying tank manifold connection contributed to the failure of the east LP-Gas tank.

5. **The incident commander should anticipate that not everything will occur in a predictable manner or in accordance with drill and training.**

The potentially defective weld was an invisible, unknown factor in the BLEVE. The LP-Gas tank's general condition will usually be an unknown to the incident commander. Deficiencies in the tank's design, construction, age, or quality of maintenance will become evident under fire and emergency conditions. Aggressive actions to cool the tank should be tempered with caution when the area around the tank can be quickly evacuated (few occupied buildings nearby) and information about the tank's condition (before and during the incident) is limited.

APPENDICES A TO E

Site Plans and LP-Gas tank diagrams

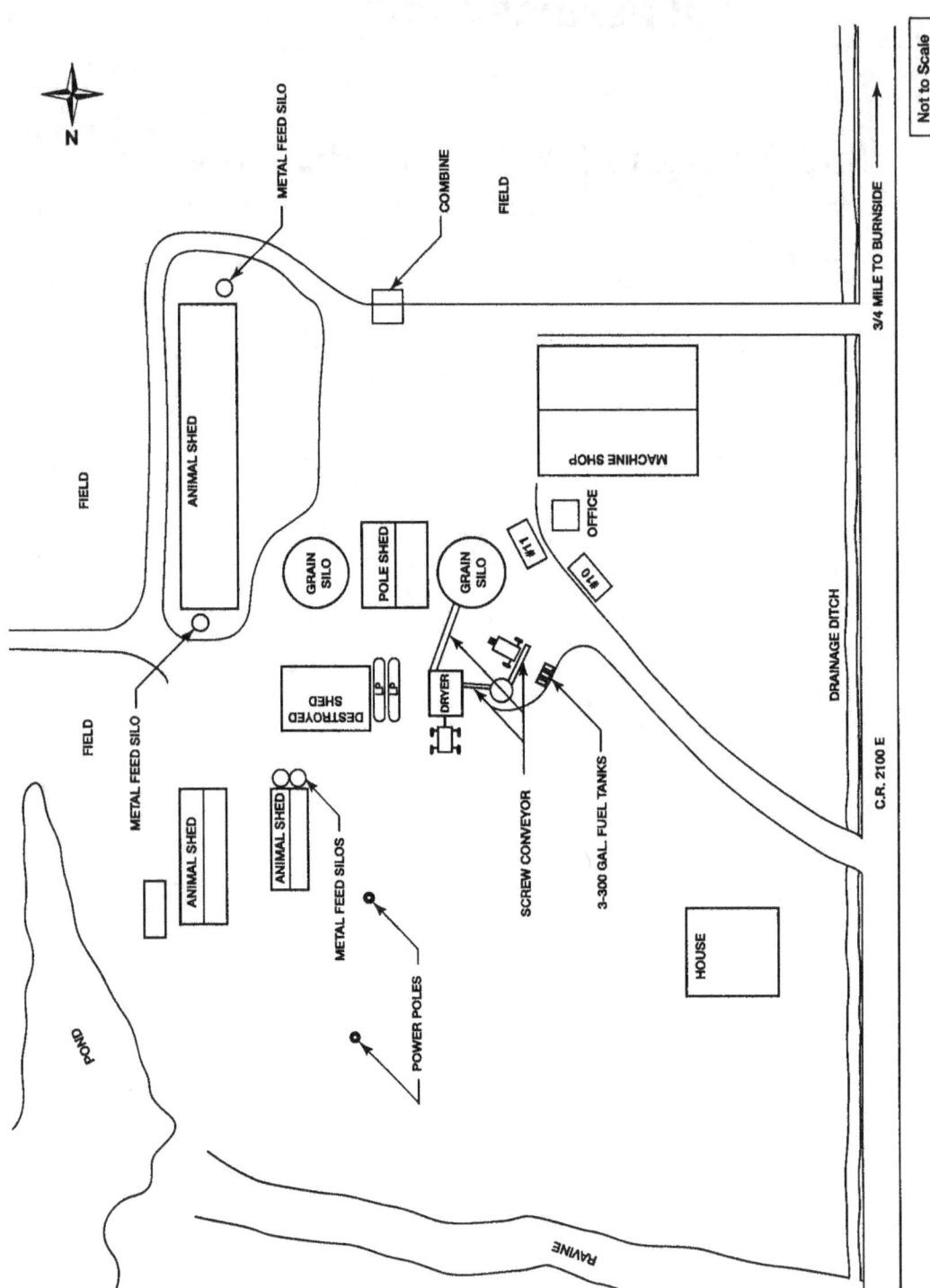

Appendix A – Site Diagram and Fire Equipment Position on Arrival

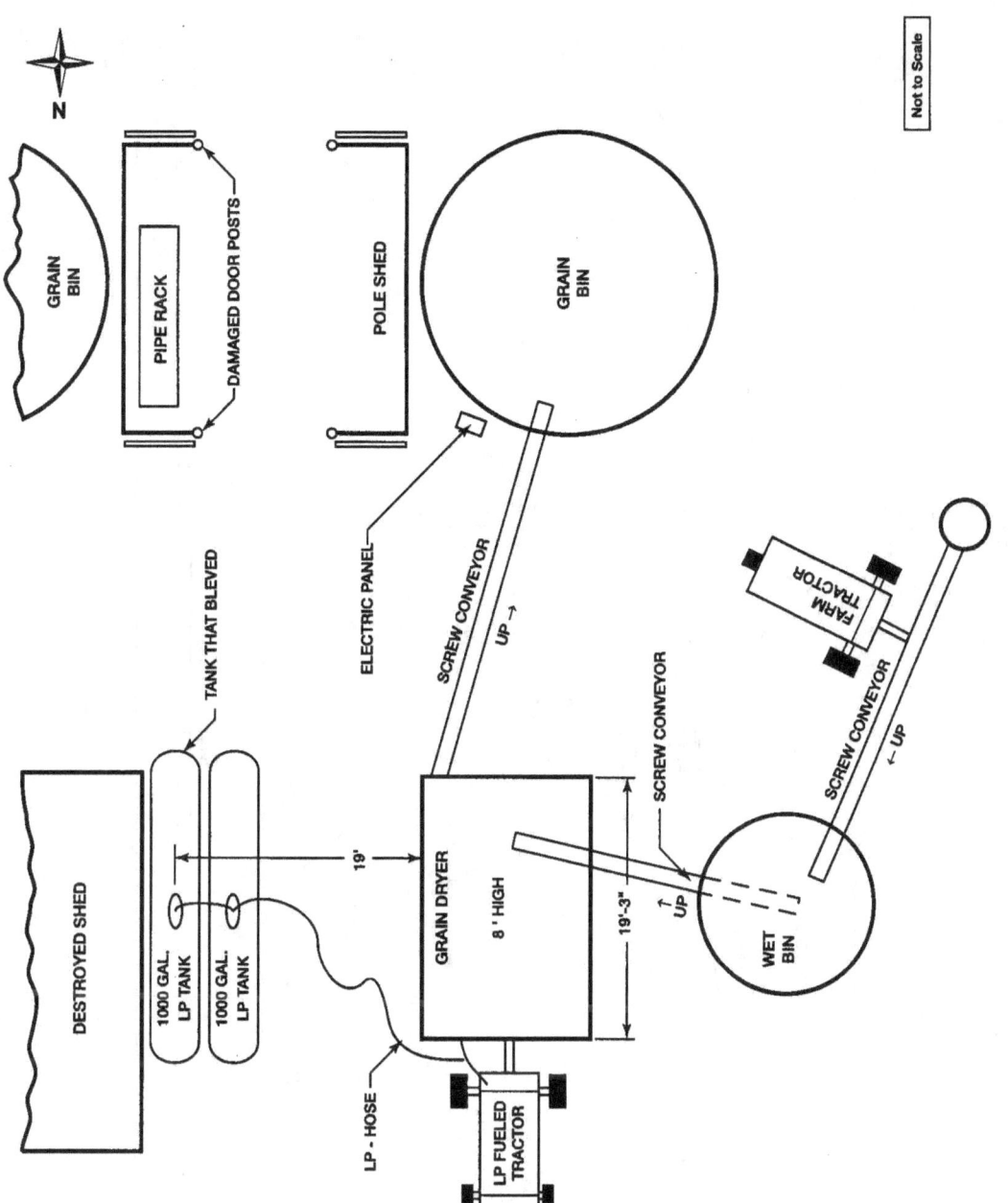

Not to Scale

Appendix B – Area Adjacent to LP-Gas Tanks

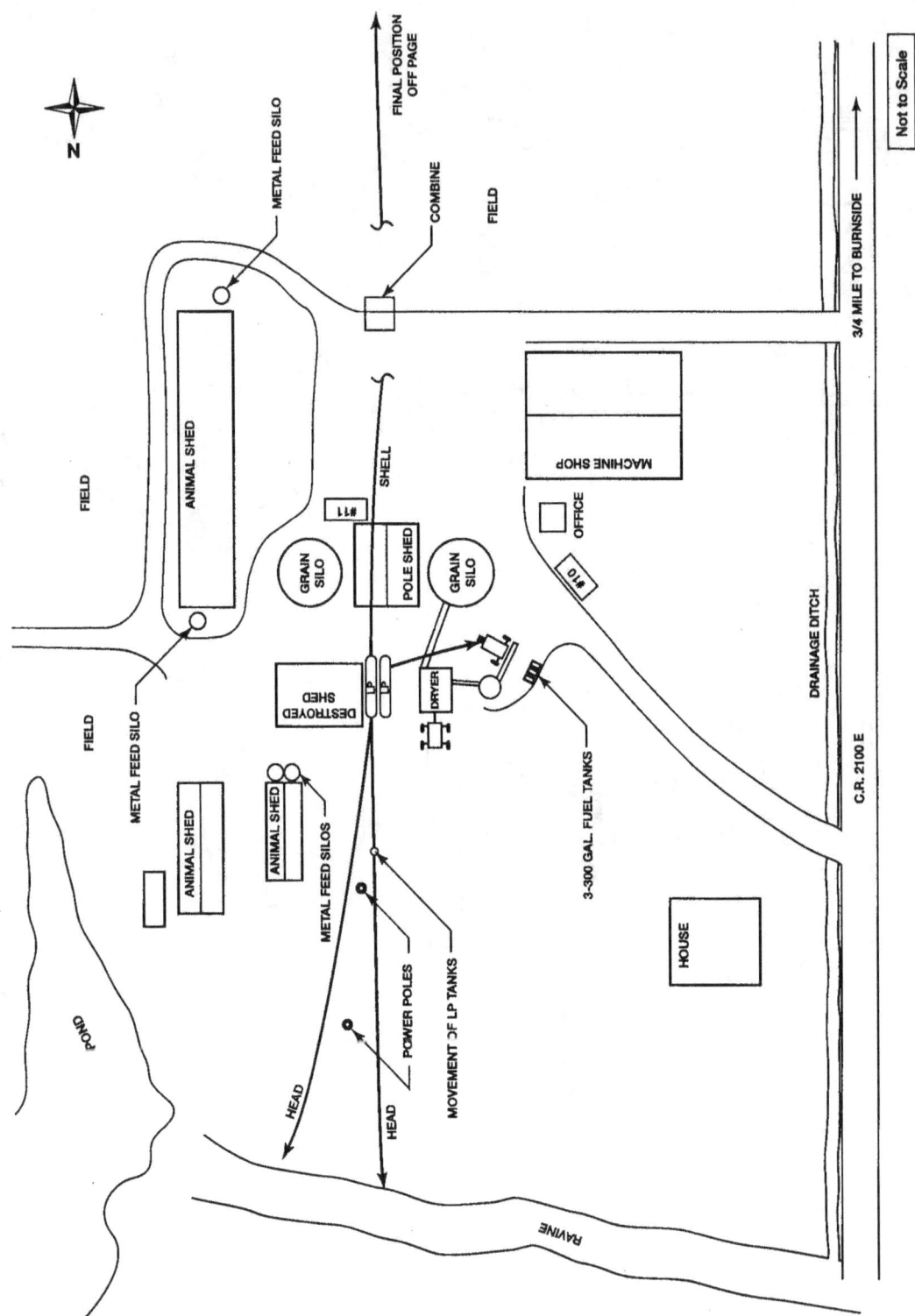

Appendix C – Fire Equipment Positions at BLEVE

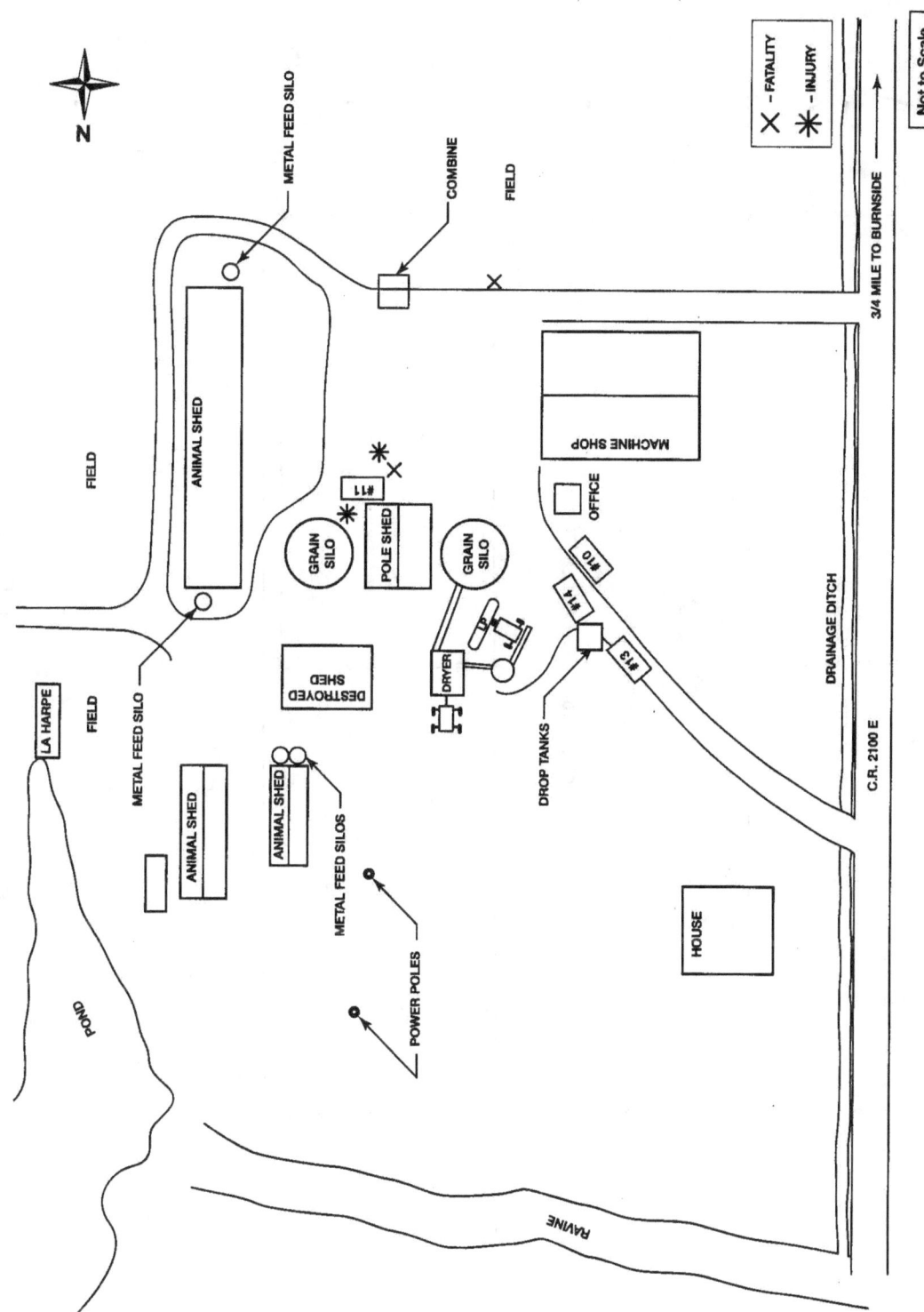

Appendix D – Fire Equipment Position After BLEVE

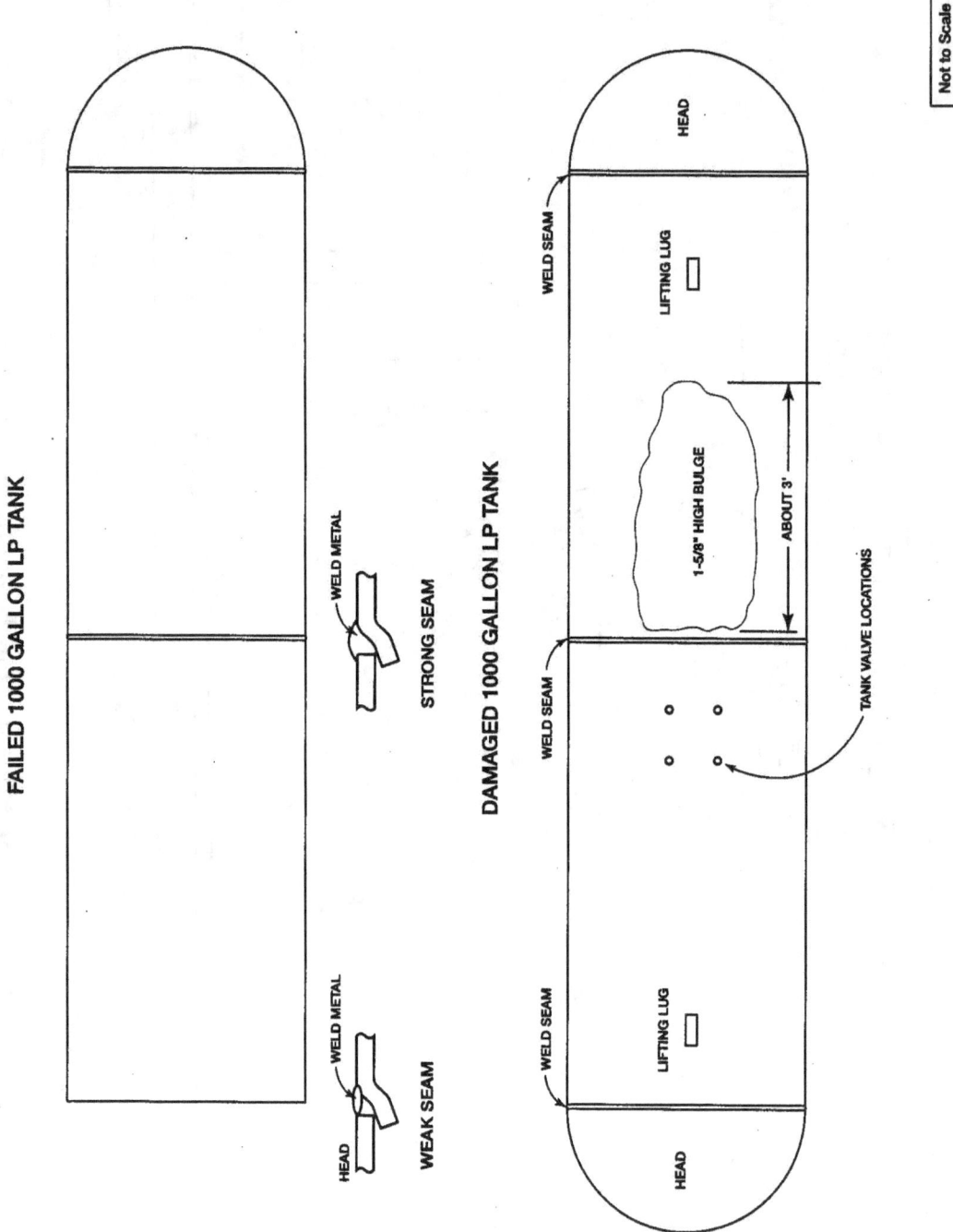

FAILED 1000 GALLON LP TANK

WELD METAL

STRONG SEAM

HEAD

WELD METAL

WEAK SEAM

DAMAGED 1000 GALLON LP TANK

WELD SEAM

LIFTING LUG

HEAD

1-5/8" HIGH BULGE

ABOUT 3'

WELD SEAM

TANK VALVE LOCATIONS

WELD SEAM

LIFTING LUG

HEAD

Not to Scale

Appendix E – LP-Gas Tank Damage

APPENDIX F

Photographs were taken by Engineering Systems Inc. and provided with the permission of the Illinois Office of State Fire Marshal.

1. Original location of LP-Gas tanks looking south. The grain dryer and LP-Gas fueled tractors are on the right. The tank passed through the pole shed in the center.

2. Concrete pad for one of the LP-Gas tanks. The pole shed's damaged 6 x 6 wood door
posts are in the background. The tank came to a rest near the house in
the far background.

3. Damage to the southeast door post of the pole shed. Three firefighters stood just on the other side.

4. Damage to the northeast door post of the pole shed. View of the grain dryer and exposures to the north.

5. The damaged grain dryer and the wood shed to the east. The screw conveyor on the right runs from the dryer's discharge to the grain silo.

6. Grain dryer, LP-Gas fueled tractor, wet bin, grain silo, and destroyed shed.

7. Destroyed field tractor that powered the screw conveyor feeding the wet bin. The west LP-Gas tank landed near this tractor on the right side of the photograph.

8. The LP-Gas tank that BLEVE'd along with the two parts of the north domed head. The protective valve cover is also in the bottom right corner.

9. The inside of the BLEVE'd tank and the rolled and welded seam between the
cylinder section and the head.

10. Close-up of a section of the rolled and welded seam between the cylinder section and the head.

11. Bulge in the metal on the top of the west LP-tank.

12. The LP-Gas hose fittings recovered after the incident.

APPENDIX G

The following data sheet was taken from the United States Fire Administration's *Hazardous Materials Guide for First Responders*. The document is available in both loose leaf and searchable CD-ROM formats. Contact the USFA at the address inside the front cover or via their Web page.

LIQUEFIED PETROLEUM GAS
(LPG)
UN 1075

Shipping Name: Liquefied petroleum gas
Other Names: Bottled gas   LPG
                Petroleum gas, liquefied

FLAMMABLE GAS 2

WARNING! · EXTREMELY FLAMMABLE!
          · CONTAINERS MAY BLEVE OR EXPLODE WHEN EXPOSED TO HEAT OR FLAMES!

Hazards:	Description:
· Odor is not a reliable indicator of the presence of toxic amounts of gas · Gas is heavier than air and will collect and stay in low areas · Gas may travel long distances to ignition sources and flashback · Gas in confined areas (e.g., tanks, sewers, buildings) may explode when exposed to fire · Contact with liquid may cause frostbite	· Colorless gas · May be shipped and stored as a compressed liquefied gas · Extremely flammable · No odor unless treated with an odorant · Floats and boils on the surface of water and is insoluble in water · Gas is heavier than air and will collect and stay in low areas

Awareness and Operational Level Training Response:	Operational Level Training Response:
· Stay upwind and uphill · Determine the extent of the problem · Isolate the area of release or fire and deny entry · Remove all ignition sources · For containers exposed to fire evacuate the area in all directions because of the risk of BLEVE · Evacuate the immediate area and downwind for a large release · Notify local health and fire officials and pollution control agencies	RELEASE, NO FIRE: · Stop the release if it can be done safely from a distance · Use large amounts of water well away from the material to disperse gas - contain runoff FIRE: · Do not extinguish the fire unless the flow of the gas can be stopped and any remaining gas is out of the line. Specially trained personnel may use fog lines to cool exposures and let the fire burn itself out · Cool exposed containers with large quantities of water from unattended equipment or remove intact containers if it can be done safely · If cooling streams are ineffective (venting sound increases in volume and pitch, tank discolors or shows any signs of deforming), withdraw immediately to a secure location

First Aid:

· Provide Basic Life Support/CPR as needed
· Decontaminate the victim as follows:
   – Inhalation - remove the victim to fresh air and give oxygen if available
   – Eye - rinse eyes with large volumes of water or saline for 15 minutes
· Seek medical attention
· Frostbite - warm injured area in very warm water
· For skin burns decontaminate with water and apply a clean dry dressing

CAS: 68476-85-7

www.ingramcontent.com/pod-product-compliance
Lightning Source LLC
Chambersburg PA
CBHW081403170526
45166CB00010B/3181